Joseph Henry Wythe

Easy Lessons in Vegetable Biology

Joseph Henry Wythe

Easy Lessons in Vegetable Biology

ISBN/EAN: 9783741183652

Manufactured in Europe, USA, Canada, Australia, Japa

Cover: Foto ©Klaus-Uwe Gerhardt /pixelio.de

Manufactured and distributed by brebook publishing software
(www.brebook.com)

Joseph Henry Wythe

Easy Lessons in Vegetable Biology

IN

VEGETABLE BIOLOGY;

OR,

OUTLINES OF PLANT LIFE.

.

BY

REV. J. H. WYTHE, M.D.,

AUTHOR OF "THE SCIENCE OF LIFE," "THE MICROSCOPIST," "AGREEMENT
OF SCIENCE AND REVELATION," ETC.

———

NEW YORK:

PHILLIPS & HUNT.

CINCINNATI:

WALDEN & STOWE.

1883.

PREFACE.

THIS work has been prepared for the students of the Chautauqua Literary and Scientific Circle. It is also adapted to use in schools.

It begins with the simplest and most elementary facts of Biology, and progresses gradually to things more intricate. It uses no unnecessary technicalities, is condensed into the smallest possible compass, and but few words are spent upon theories. The Christian philosophy of life, which the author believes to be the only true philosophy, is clearly stated; but he aims to present only facts, and the plain inferences from facts, to the attention of the inquirer.

CONTENTS.

EASY LESSONS

IN

VEGETABLE BIOLOGY.

CHAPTER I.

WHAT BIOLOGY TEACHES.

1. THE word Biology is made up of two Greek words—*bios*, life, and *logos*, a discourse. It means the study of living things.

A few years ago all the living and non-living objects in the earth—minerals, plants, and animals—were embraced in one department of knowledge, called Natural History, but it is now usual to study living and non-living objects in different departments. A knowledge of Biology involves much more than the ability to distinguish the different kinds of living beings so as to be able to label dead specimens in a cabinet. This science uses the terms of Botany and Zoology only as the builder uses a scaffolding for the erection of an edifice, or as a postmaster the boxes of his office for a more convenient classification.

2. Biology includes in its survey both animals and vegetables, and considers their forms and peculiarities,

the parts of which they are composed, their relations
to each other, and the uses which they serve. It takes
in the entire life-history of every living thing, with
the changes which occur in health and disease. If
fully recorded, the world would hardly hold the books
which might be written about these things, since there
are many thousands of species, or different kinds,
both animal and vegetable; and the influences to
which they are subject are quite innumerable. Yet
many observations and comparisons have shown that
one kind agrees with another in certain particulars,
so that the general principles which underlie their
forms, changes, structure, and uses may be under-
stood. The consideration of these general principles
and correspondences makes up the chief part of the
study of Biology.

3. Biology studies only living beings. The general
forces of nature and the changes in non-living matter
are the subjects of Physics and Chemistry. Biology
only refers to these changes as they affect living
things, or are modified by the presence of life.

4. Those astronomers who still hold to the Nebular
hypothesis—the theory that the sun and planets developed themselves out of a sort of fiery vapor or
nebulous matter—teach us that only a few of the
planets of our solar system are capable of sustaining
life, and chemical analysis shows that only four out
of nearly seventy elementary or simple substances, of

which the world is composed, are found essentially connected with living beings. Geology also shows us that there were vast periods in the past history of the earth when no life existed. These periods are hence called *azoic*, or without life. At the present time, the ice-fields of the poles, and the great deserts, are to a large extent lifeless. It is plain, therefore, that life is not an essential part of creation. Suns and planets would shine; gravity, light, heat, and electricity would operate; and chemical changes take place, if there were no living beings. From this consideration we see clearly that Biology embraces something more than the study of the laws and phenomena of non-living matter. It is the science of life. It concerns itself with every thing pertaining to life.

5. The question, What is life? has given rise to various speculations in every age of history to the present day. Some of the Greek philosophers taught that it was the result of the harmony or agreement of the different parts of the body; and this view is repeated in modern times by those who claim that life is the result of organization. It requires but little thought to see that this is no explanation at all. Since all organization depends on living matter, as we shall more fully perceive hereafter, one might as wisely say that an architect or carpenter was the result of a house, as to say that life results from organization.

Some have claimed that life is a sort of refined

matter; a gas, or ethereal vapor. Heat, oxygen, and galvanism have all had their advocates, and at the present day, when teachers of physical science show that one kind of force (as motion, light, heat, or electricity) can be converted, or changed, into another, some are found to say that physical force can be changed into life, or vital force. But this theory is insufficient to explain what life is, since it does not show what is meant by "physical force," nor what changes its form. Life must either be the power of matter or spirit, for we can only conceive of these two forms of existence. This theory is also insufficient, since it does not account for the living germ through which (according to the theory) heat or other form of force passes in order to be converted into vitality.

6. Writers who have endeavored to define life in accordance with the teaching of materialism—which makes matter explain every thing—have involved themselves in unsatisfactory and often absurd conclusions. The teaching of the Bible and of all the religions of mankind, the belief of the most eminent philosophers, the doctrine held by the early Christian fathers, and maintained by the majority of scientific and unscientific men, is that the difference between a living body and the same body after death arises from the union of matter and spirit. In other words, a living thing is a spiritual essence which clothes itself with material particles after a form and according to

an order (or law) of its own kind. Dr. Noah Porter, President of Yale College, in his work on the "Human Intellect," devotes an entire chapter to prove that life and soul are synonymous words, and apply to all living things. The writer of the present work, in the volume entitled "The Science of Life," regards life, not as identical with soul, but as the sum of the activities resulting from the union of mind and matter. Those who desire to study the subject more thoroughly are referred to those works.

CHAPTER II.

LIVING MATTER.

1. A CAREFUL study of any living thing, either vegetable or animal, will show that it is not all alive. Some parts are dead, although they may retain the form impressed on them by vitality, as in the case of a dry branch of a tree, and may serve various purposes in relation to the rest of the body, like resin in some plants and milk in animals. When we clip the hair, or pare the nails, we cut merely the dead, insensitive part; but at the root of the hair or nail we find the "quick"—the living part. By the use of the microscope it has been found that every part of every living body, or organism, has scattered through it little particles of living matter. In the skin, flesh, bones, and nerves of animals, and in all the different parts of vegetables, these small particles of living matter may be seen by a good instrument. It is the presence of this living matter which entitles any thing to be called a living being. Without this a man, a horse, a tree, or a flower, would be as dead as a piece of iron or chalk.

2. This living matter, seen through the microscope, looks like a bit of jelly or albumen. It is generally

transparent, and is neither quite solid nor fluid. When it was first discovered it was thought to be inclosed in a sort of membrane like a bladder, and it was called a *cell*. It is now known that it has not always an outside membrane. It is often called *protoplasm*, or first formation. It is also called by the better term *bioplasm*, or living formation. Some recent discoveries with the microscope render it likely that the real living matter in each cell, or piece of bioplasm, is arranged like a net-work, and communicates with neighboring cells so as to make a continuous living structure throughout the body, either of plant or animal.

Those who have access to a good microscope may find an example of this living matter, or bioplasm, in the white blood-cell. Prick your finger, and put a small drop of blood, about the size of a full "stop" —printer's type—upon the thin glass cover of a microscopic slide, then quickly put the cover in its place, so that the drop may spread by capillary attraction, and observe the globules in the field of view of the instrument. In about every three hundred of the ordinary red blood-disks you will see one of the white disks. If you keep it warm by a heated stage, or if you examine the blood of a cold-blooded animal, as a frog, instead of your own, you will be able to see its peculiar motions and other phenomena.

3. From such simple, jelly-like particles all animals

and vegetables originate, and by such particles are all organic structures built up. Bone, muscle, nerve, and skin, in animals, and fiber, wood, and vessels, in vegetables, are all constructed from such elements. It is the province of biology to find out how this is done.

4. The particles of bioplasm, or living matter, always look alike, no matter where they belong. There is no difference under the microscope between the bioplasm of a blade of grass or a whale, of an oak, a rose, a dog, or a man. The bioplasm of skin cannot be distinguished from that of flesh, or of the blood, yet there is a wonderful difference in the power and products of these different kinds, although the difference is not visible even with a microscope.

Chemical examination also shows that all living matter is composed of the same elementary materials. Oxygen, hydrogen, carbon, and nitrogen enter into the construction of every piece of bioplasm. Sometimes lime, or iron, or sulphur are also found, but these are accidental, and not essential or necessary.

5. The jelly-like living matter which has been described, like any other jelly, is permeated or saturated with fluid, which may be considered apart from the rest of the structure. All organic substances contain considerable water. A human body weighing a hundred and fifty pounds can be dried in an oven until it weighs only seven and a half pounds. The water in a piece of bioplasm may also contain other

substances in solution which may serve the living matter as food. In the spaces between the minutest particles or net-work of bioplasm we may also find material which has been formed by it. So that in every bioplast, or living particle, we recognize matter in three different states : (1) Matter not yet alive, but about to become so, called *Pabulum*, or nutriment. (2) Living matter in the strictest sense, or *Bioplasm*. (3) *Formed material*, or matter which was alive, but is so no longer.

Owing to the constant action of the air and other influences, the formed material is constantly decaying, or becoming effete, and thus returns to the inorganic world from which it originated, so that we may say of the body of any living thing—though not of the life—"Dust thou art, and unto dust shalt thou return."

6. Physical forces, like gravity, heat, light, and electricity, and chemical agencies, affect all living matter as well as the non-living, but not always in the same manner. Some forces, as gravity, for example, act in the same way on the living and the non-living. The passage of liquid out of or into a porous or membranous vessel is similar in a living or non-living thing. In other instances the presence of life greatly modifies the influence of inorganic force. Thus, water generally freezes at 32 deg. F., and boils at 212 deg.; but bioplasm, or living matter, resists the

extremes of heat and cold as long as the life remains. There are differences in this respect in different organisms. The motions of some simple forms of animals are arrested by ice-water, and recommence on increasing the temperature, but the development of trout's eggs proceeds well in ice-water, while in a warm room they soon die. Many kinds adapt themselves to a considerable change of temperature if it be gradual. Men will endure the cold of an Arctic winter, and, on the other hand, the workers in plaster will bear for a considerable time the heat of an oven raised to 500 deg. F. The influence of light, heat, and electricity upon bioplasm varies according to the species, some requiring a greater amount than others for natural growth. As to chemical changes, the life power has greater modifying power than any thing else in nature. A vast variety of products in both animals and vegetables are due to the controlling influence of life. In a few instances the skill of a chemist has imitated the product in his laboratory, but the majority can only be found as the result of life. Among them may be named albumen, starch, sugar, and gum. Many others will occur to us as we proceed in our studies.*

* See "Science of Life," Chap. IV.

CHAPTER III.

DIFFERENCES BETWEEN LIVING AND NON-LIVING MATTER.

1. THE jelly-like bioplasm described as existing in all kinds of vegetables and animals, and forming all sorts of organic materials, is so different in power from non-living matter as to compel us to believe that it contains something more than mere matter. Each kind, or species, of living being can do something which is peculiar to itself, yet there are certain particulars in which all kinds agree, or things which all sorts of bioplasm can do, but which no matter can do which is not alive.

2. All bioplasm has spontaneous motion. Most vegetables are fixed to one spot, but the living matter in their tissues is just as much in motion as the bioplasm of the most active animals, so that the same things may be said of either animal or vegetable bioplasm. Non-living matter is passive. It has *inertia.** It can neither originate, suspend, nor destroy motion. It can only transmit motion, or be moved. But bioplasm, or living matter, has primary energy, and can overcome inertia. Its motions are spontaneous, or

* A property of matter which causes it to remain in a state of rest or of motion.

2

spring from its own internal energy. So far from being caused by external influence, its movements are often in direct opposition to gravity, or any other force which we may imagine to act upon it.

The motions of bioplasm are of three kinds:

(1) Inherent motions of the individual particles among themselves. Each particle is as much alive as the whole mass, and the movements of each are spontaneous. If a thread or filament of bioplasm be examined in a powerful microscope, the motions of the particles may be observed by the granules of formed material (Sec. 5, Chap. II) which may be scattered through the mass. " As the passengers in a crowded street may go the full length of the street, or turn back, or stop and double, as many times as they wish, so do the particles move in the mass of bioplasm. Up, down, across, backward, and in all directions— even through each other—do these molecules move, each impelled by its own inherent energy." *

Observe another motion of bioplasm:

(2) Constant change of shape. A piece of bioplasm never remains at rest. If unconfined, its external appearance or shape is continually changing. This has been called *amœboid motion,* from the name given to one of the simplest forms of living beings known, the *Amœba.*

Fig. 1. This appears to be simply a piece of bio-

* " Science of Life."

plasm, or living jelly, yet it is a complete organism. Its organs, however, are all extemporaneous. It never retains the same outline or form, and it can project any part of its substance in the shape of an

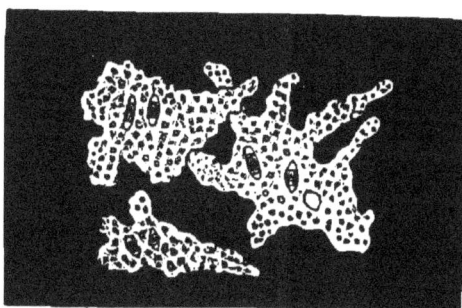

Fig. 1.

arm or branch, and if it touches any thing which may serve it as food, the rest of the body will flow around it and digest it. Indigestible parts it discards by flowing away from them. It literally swallows without a mouth, digests without a stomach, and moves without muscles, while a small fragment of its substance is capable of repeating the same things as an independent organism. Such living Amœbæ are found in stagnant pools of water and many other places. Since every piece of bioplasm has similar changes of shape, it is agreed to call such changes *amœboid*.

(3) Wandering movements. As unconfined bio-plasm can flow along an arm or branch of its own body around its food, it can in the same manner change its position from place to place. In this way

the white blood-cells, described in Sec. 2, Chap. II, wander out of the blood-vessels in order to construct the various parts of the body as they may be needed. (Fig. 2.)

3. Another peculiarity of living matter is the power of nutrition and growth. The non-living increases in size by external additions, but bioplasm selects appropriate material from its food, (or pabulum,) changes the chemical relations of this material,

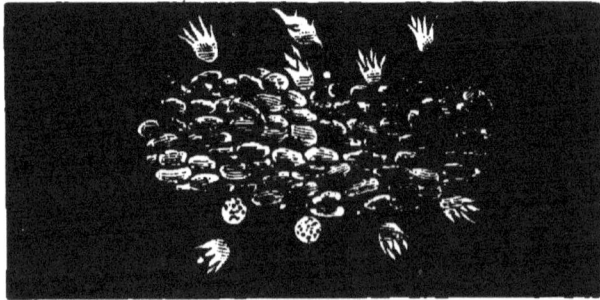

Fig. 2.

and appropriates it to its own structure in such a way that it grows from within. It is altogether different from the enlargement of a crystal, or of the increase of any thing without life.

4. Bioplasm can generate or reproduce its own kind of living matter. No living being exists which did not originate from living matter of a similar kind. Some persons have thought that they have seen the beginning of life in some forms of simple beings, originating by what has been called sponta-

neous generation, but a careful examination proves them to have been mistaken. Microscopic living beings appear after a time in water where vegetable or animal matter has been left to decay, but we now know that their eggs or seeds, in the shape of very minute particles of bioplasm, are to be found in great numbers floating about in the air and in every collection of water, and that these seeds or eggs retain their life-powers after having been subjected to boiling heat. Of course there was a first animal or vegetable of each kind, and some learned men think it was the offspring of another kind, not quite like it, by some sort of change of form or of activity, so that the more complicate forms sprang from more simple ones. This doctrine of "evolution," as it is called, has never been proved by facts, although the varieties existing in the same species, as the different kinds of dogs or pigeons, give it a sort of probability to some persons. All kinds of living beings, however, spring from similar parents, and none have ever been known to change into other forms. The forms pictured upon the monuments of early Egyptian history are similar to those of the present day. New varieties, not new kinds, may be produced by culture, as in the case of roses and other flowers, but such will return to their original form if allowed to grow wild.

5. The power of a living thing to preserve its own identity amid all the material changes which take

place, is entirely different from the power of mere matter. In Chap. II we learned that oxygen, hydrogen, carbon, and nitrogen are found in every particle of bioplasm, and that sometimes a few other chemical elements occur accidentally. Now these things do not remain quiet in living matter. Atom by atom they quickly pass through it. They are seized by the bioplasm as food, transformed into its own structure, and then are changed into formed material, as starch, wood, gum, oil, etc., in vegetables, and blood, muscle, bone, and nerve in animals. The formed material decays atom by atom and is cast off, to mingle again with the inorganic elements of the world. During all these changes the living being preserves its identity and power. Thus it is possible that an atom of oxygen or hydrogen may be cast off from some of the bioplasts of our own bodies, be wafted by the air to the sides of the Andes, be appropriated to the use of the bioplasm in one of the cinchona-trees, and return to us in the form of quinine, perhaps to cure us of ague. The possibilities of science exceed the most romantic imagination. The preservation of its identity shows that the life of bioplasm differs from the atoms which come to and go from it. It does not depend upon the new ones, for it existed without them, nor upon the old ones, for it remains without them. Life is not matter, but matter's master.

CHAPTER IV.

DIFFERENT KINDS OF LIVING MATTER.

1. WE have learned in former lessons that the elementary particles of living matter look alike and have the same essential composition, although they may belong to different kinds of living beings, and may serve very different purposes.

2. Different bioplasts produce different forms of living things by the special instincts (or tendencies) belonging to each kind. These different forms are arranged, or classified, in Botany and Zoology, not only for convenience of study, but as far as possible in accordance with the plan of Creative Wisdom, which has assigned to each its place in the order of nature.

3. The thousands of different kinds of vegetables and animals could not be remembered; but by grouping certain kinds together, which are in some respects similar, and by combining these groups with others, naturalists are able to form something like an orderly system. Such a system will be a natural system if the grouping be according to existing resemblances or true affinities, otherwise it will be an artificial system. An illustration of a natural system of classification may be

found in a house, considered as a building. Let us suppose that all kinds of houses may be referred to one or other of three types, or general plans—the Oriental type with a dome roof, the Grecian type with a flat roof, or the Gothic type with a pointed roof. We select a house of the Grecian type. But there are several classes of the same type, as made of wood, iron, or stone. Our supposed house is a wooden one. But there are several orders in each class. Thus we may have Pine, Oak, Mahogany, etc. The house we are considering is built of oak. In each order there are other groups, or genera. The doors may be at the front or at the side. Of the kind, or genus, having the door at the front, there are more specific kinds, according to the resident. In our supposed case, the specific house is John Smith's. Thus, if we have rightly classified we have learned that John Smith's kind, or *species*, of house, is of the *genus*, or group, which has front doors belonging to the larger group, or *order*, of houses built of oak. This again is grouped in the *class* of wooden houses under the Grecian *type*.

In a similar manner naturalists try to group together living things according to their real relationships. All systems, however, are imperfect, on account of our imperfect knowledge, and we need not be surprised to find one naturalist referring a form to a group very different from that to which it is referred in the scheme of another, since men differ greatly, not only in knowledge, but in opinion.

Types represent general plans of structure. *Classes* are formed by the special modification of a type. *Orders* are groups of the same class related by a common structure. A *Family* or *Genus* is a still smaller group having generally the same essential structure. A *Species* is the smallest group whose structure is constant. Species are so much alike that they may have descended from the same parents. *Individuals* are the units of organic life, forming a complete animate existence. Peculiarities of races or breeds are called *varieties*. Vegetables and animals are distinguished from each other by the term *Kingdom*, and the types in each kingdom are called *Sub-kingdoms*.

4. The names which naturalists use to designate kinds or groups of living things are formed from the Latin or the Greek language. There are good reasons for this. In the modern languages common things have many different names. But the majority of plants and animals are new or rare, and have no name in any modern language. If there must be a new name it may as well be Latin or Greek as any other. Then the latter languages have the advantage of expressing by their combinations some peculiarity which is distinctive, and which is readily recognized by the learned. As to the difficulty of learning these terms, it is only apparent. They can easily be mastered by

application. The terms of natural science are no more difficult than those of geography or of history.

5. The types, or general plans of structure, found in vegetable Biology may be described briefly as follows:

1.) PROTOPHYTES, (Greek *protos*, first, and *phuton*, a plant.) The first or simplest forms of plants—vegetables composed of a single cell, or mass of bioplasm.

2.) THALLOGENS, (Gr. *thallus*, a frond, or vegetable expansion, and *ginomai*, to produce or grow.) Plants composed of a tissue of cells, or bioplasts, but with no clear distinction of stem, root, and leaves.

3.) ACROGENS, (Gr. *akra*, summit, and *ginomai*, to grow.) Plants which grow at the summit only, and not in diameter.

4.) ENDOGENS, (Gr. *endon*, within, and *ginomai*, to grow.) Plants whose vessels and woody fibers first grow within the stem. The seed has but a single lobe, or cotyledon.

5.) EXOGENS, (Gr. *exo*, outward, and *ginomai*, to grow.) Plants whose woody fibers grow in outer layers. The seed has two lobes, or cotyledons.

Under these five types or plans of structure all the multitudes of plants which clothe the earth or dwell in the sea can be arranged.

CHAPTER V.

INDIVIDUAL VEGETABLE CELLS.

1. THE elementary masses of bioplasm (Sec. 2, Chap. II) are usually called cells, even if they are merely pieces of animated jelly, uninclosed by an outside shell or membrane. Some of these cells remain unconnected with others during their entire life, and multiply by self-division. Others have a living or vital connection with neighboring cells, so as to form tissues and organs. In this chapter we shall consider the vegetable cell as an individual—its various appearances and its activities.

2. It is not easy to distinguish between the cell, or living matter, of an animal and a vegetable. Some cells appear like animals at one part of their lives and like vegetables at another part. After long study learned naturalists have agreed that the principal difference between animals and plants is that the latter can be nourished by simple mineral or chemical (that is, unorganized) matter, while animal nutrition requires material which has been organized, or made part of a living being.

3. Most vegetable cells produce a membrane, or *cell-wall*, on the outside, within which the living matter

is, as it were, imprisoned, although certain openings, or pores, may be left in the cell-wall for the purpose of communication. There is also a concentration of living matter within the cell, called a *nucleus*, and sometimes a still further concentration within the nucleus, called a *nucleolus*, (or little nucleus.) The bioplasm, also, within the cell, differs in density, that next to the wall of the cell being thicker, or more mucilaginous, than the rest. This latter part, or layer, has been called the *primordial utricle*, or original bag. Fig. 3 will give a general idea of the elementary vegetable cell.

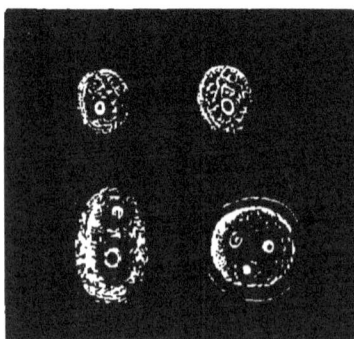

Fig. 3.

4. The cell-wall referred to in the last paragraph is composed of a substance somewhat like starch, called *Cellulose*. This is often thickened by deposits inside, layer after layer. When it becomes solid it is known as woody tissue. Common wood is made up of a number of these cells arranged side by side.

5. In cells with thin membranes, the inherent motion of the bioplasm (Chap. III, Sec. 2) can often be seen with the microscope. This motion has been called circulation, but it is an irregular motion of the particles, sometimes slower, sometimes advancing, now

retreating, now stopping, or beginning again, in a way differing from all non-living matter. (Fig, 4.)

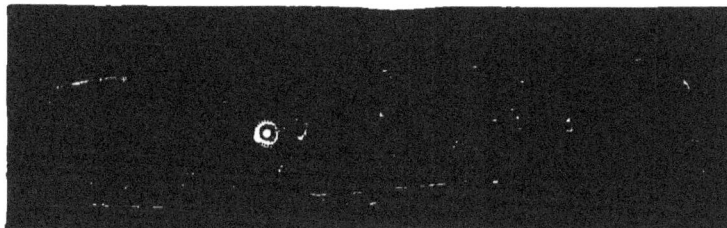

Fig. 4.

6. Sometimes the cell-wall has a deposit of flinty material, (*silica*,) or of other mineral matter, which is often beautifully marked with lines and dots. Different species of plants produce different patterns of these deposits, as represented in the figures of Diatoms in Chap. VII.

7. The cell-wall is often irregularly thickened by a deposit inside, so as to present different appearances in different cells. In the Pine and Fir tribe the pores in the wall of the wood-cells are surrounded by concave spaces or depressions. (Fig. 5.)

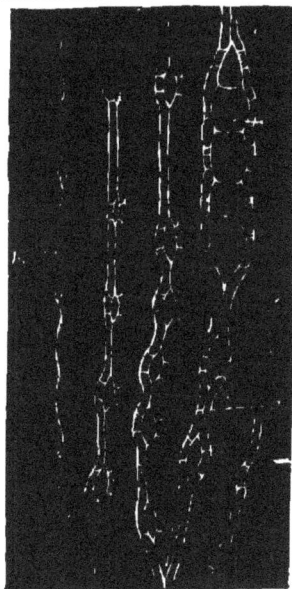

Fig. 5.

In other cases the more solid matter is deposited so as to form dots, or rings, or spiral fibers, on the cell-wall. (Figs. 6 and 7.)

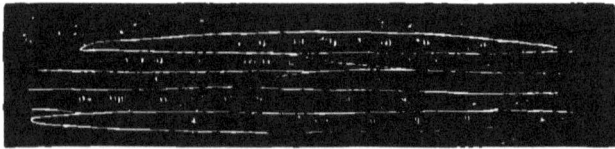

Fig. 6.

8. Vegetable cells are of various shapes, according to the purposes which they serve or the pressure to

Fig. 7.

which they have been subjected. They may be globular, oval, conical, prismatic, cylindrical, branched,

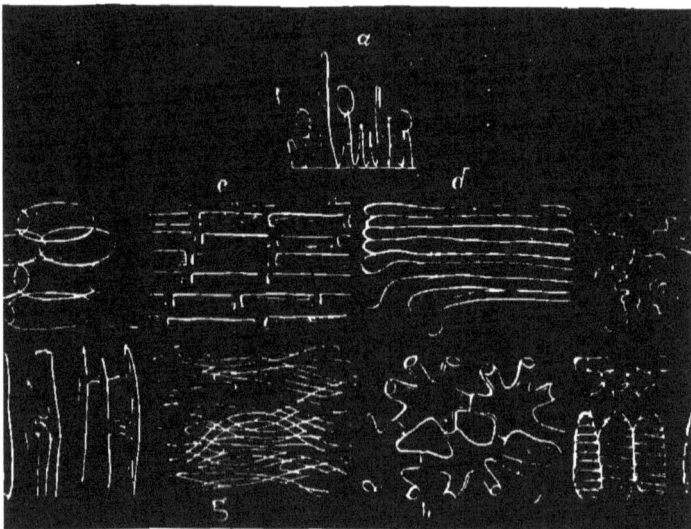

FIG. 8.—Various forms of cells: *a*. Conical. *b*. Oval. *c*. Prismatic. *d*. Cylindric. *e*. Sinuous. *f*. Branched. *g*. Entangled. *h*. Stellate. *i*. Fibro-cellular tissue.

star-shaped, hour-glass shaped, disk-shaped, tubular, many-sided, or of any other form. (Fig. 8.)

9. The bioplasm within the cell-wall may be transformed into a great variety of formed material, making special *cell-contents.* These may be solid, as coloring matter, starch, crystals, and resin; or fluid, as oil and gum, or solutions of sugar or tannin. The most important of these substances is called *Chlorophyll,* (Gr. *chloros,* green; *phyllon,* a leaf,) or the source of the green color of plants. It is composed of a peculiar coloring matter intimately united with separate particles of bioplasm, which, under the influence of sunlight, causes the absorption of carbonic acid gas from the air, which is necessary to nourish the plant. *Starch* is also an important product of vegetable cells, even more widely distributed than chlorophyll. It seems to be stored up in the cells as a reserve food-material for the use of the new cells which are subsequently formed, hence it occurs in large quantities in seeds, bulbs, and tubers. (Fig. 9.)

Crystals of oxalate of lime often occur in cells, as well as *acid substances* and *alkaloids,* like strychnine, quinine, etc., dissolved in the cell-sap.

It is remarkable that such different materials as cellulose,

Fig. 9.

chlorophyll, starch, gum, resin, oil, etc., may be produced from similar cells under the influence of the same environment, and equally exposed to heat, moisture, and electricity. (Fig. 10.)

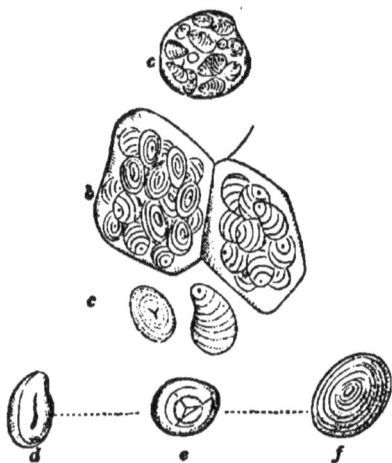

10. Cells generate by self-multiplication. One will divide into two or more pieces of bioplasm, which will assume the form and function of the original cell in the simpler forms of plants, or may take a different shape and use in the plants which are composed of numerous cells.

FIG. 10.—*a, b*. Cells of a potato, containing starch. *c*. Starch-grains apart. *d, e, f*. Wheat-starch in different positions.

If the *mother-cell*, as it is called, possesses a nucleus, the self-division is preceded by the formation of new nuclei, one for each of the *daughter-cells*. Sometimes the self-division of the bioplasm is produced by the projection of a sort of bud, which is separated from the parent mass. If the newly-formed cells retain a vital connection with each other, cellular structures, or tissues, of various sorts are produced. The most complete development of this kind occurs in the higher types of plants.

CHAPTER VI.

THE VEGETABLE CELL AS A MEMBER OF A GROUP.

1. ONLY a comparatively small number of plants consist of a single cell. These are the simplest forms of plant life. In the greater number of vegetables the cells are united into groups. Cell-families may originate from a single mother-cell, and remain for a time closely connected, or contiguous, yet each daughter-cell preserves its own individuality, and may originate a new colony. Such cell-families only occur in the lowest classes of plants. In the higher classes the union of cells which forms tissues and organs is permanent, and the separate cells are often so closely united as to form a single cavity or vessel; the cell-walls of young contiguous cells fusing into a common mass, or cells originally distinct uniting in those parts of the walls which are in contact. This union is so strong as only to be destroyed by chemical agents which dissolve the cell-wall. Yet sometimes cells which have partially united separate from each other, forming a cavity. These spaces also may be grouped together so as to form passages, or air-canals.

2. The woody fibers of plants, and the cellular tissue which makes the softer, fleshy, and pithy parts,

3

are made by the union of cells into groups. Observation has shown that in the higher plants new cells are not produced every-where uniformly, but in particular spots. To places of this kind the terms *growing-point* and *growing,* or *formative, layer* have been applied. Growing-points may be seen in the tips of buds, and formative layers between the wood and bark of trees. The names *formative* or *generating tissue* have been given to the tissue which is here formed by the division and union of cells. A tissue in which the cells are not capable of self-division is called a *permanent tissue.*

3. In direct contrast to the generating tissues are the *healing tissues,* or cork tissues. In these the cells lose their cell-sap, and the cellulose of the walls becomes converted into cork, which is of great importance as the true healing tissue of plants. This is formed like a cushion, or callus, over the surface of a wound in a tree. The cuttings of the cochineal cactus would decay at once if set in the ground with the surface of the wounds fresh. They are therefore laid for some time in the sun, in order that a cork tissue may be formed, which closes the wound and prevents decay. Such facts not only prove that the living vegetable is governed by other than invariable mechanical forces, but are also mute prophecies of higher truths revealed to the human intelligence in God's word relative to the healing of the soul.

4. *Vessels* are made by the union of several cells, the partition-walls disappearing while the union continues at the margin. Such vessels may be dotted, reticulated, annular, or spiral, from the deposit on the cell-wall. Chap. V, Sec. 7. (See Fig. 11.)

Bast-tubes, or *bast-fibers*, are long, pointed, thick-walled tubes, commonly united into bundles. In hemp, flax, etc., they form textile fibers, and they are sometimes united in the inner layer of bark so as to form a kind of lace, as in the lace-bark of the West

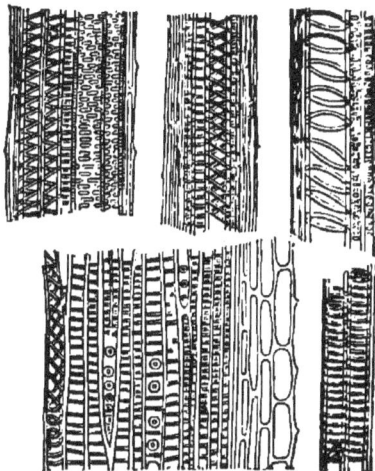

Fig. 11.

Indies. *Sieve-tubes*, or *bast-vessels*, result from the joining of cells standing one above the other, the partition-walls of which have become perforated. Some have sieve-like perforations through their side walls.

Other vessels are simple or branched tubes, often making a net-work, and containing a sort of milky fluid called *latex*. This latter contains different substances in different plants, as gum, resin, opium, india-rubber, etc.

5. In addition to the groups of cells which form

tissues and vessels, other smaller groups are found, with receptacles formed by passages between the cells. According to · the nature of the substance secreted these spaces are called *oil-passages*, *resin-passages*, *camphor-glands*, *resin-glands*, etc. The term *nectaries*, or honey-glands, is given to any part of a flower which secretes honey or sugary fluid.

6. The first independent tissue formed in flowering plants by the union of cells is the *epidermis* or skin. The outer layer of epidermal cells is transformed into a thin, structureless membrane called the *cuticle*. The form of the cells varies in different plants, but they are usually flat or tubular, but sometimes projecting like little knobs or bladders, which gives a velvety or glistening appearance to the leaf or flower.

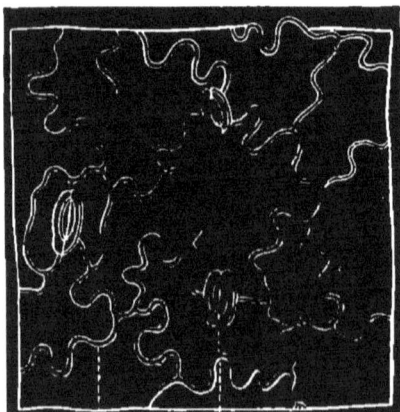

Fig. 12.

Among the epidermic cells are found pores, each of which is called a *stoma*, or mouth. (Plural, *stomata*.) These are inclosed by two or four cells, which are crescent-shaped, and distinguished from other epidermal cells by their smaller size and by containing chlorophyll.

These cells are thought to regulate evaporation by their expansion. (Fig. 12.)

Hairs are epidermal structures, composed of one or more cells. Under this general term may be included prickles, scales, stinging-hairs, and glands. Some secrete volatile oil, and others, as the nettle, an acrid fluid. Fig. 13 exhibits some of their forms.

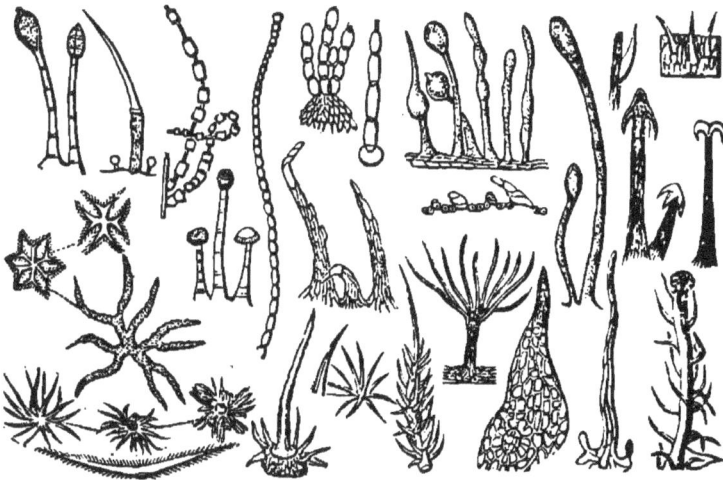

Fig. 13.

7. Next to the epidermis we find the *cortex*, or bark, often composed of cells containing starch or chlorophyll. Vessels containing latex (Sec. 4) and glands, as well as sap-passages between cells, may also occur in it. In some plants, masses of cork may be found in the bark or beneath it. In such the outer parts die and the bark peels off.

8. Beneath the bark is the formative layer or

cambium, (Sec. 2,) in which thin-walled cells become transformed into vascular or bast-cells, (Sec. 4,) and these are changed into permanent cells. Groups of cells are thus formed which, united into bundles, penetrate the rest of the tissue, forming the *fibro-vascular bundles*. The development of these bundles is characteristic of different types of plants. The simpler types have no fibro-vascular bundles, and are called *Cellular Plants;* the rest are termed *Vascular Plants*.

9. The *fundamental tissue* generally consists of thin-walled cells containing starch, although other forms of cells may be present. In plants which have no fibro-vascular bundles the whole interior may be regarded as fundamental tissue. In other plants it fills up the spaces between the bundles and within the bark. In the type of Endogens (Ch. IV, Sec. 5) this tissue is most developed, while in Exogens it occupies a smaller portion of the structure.

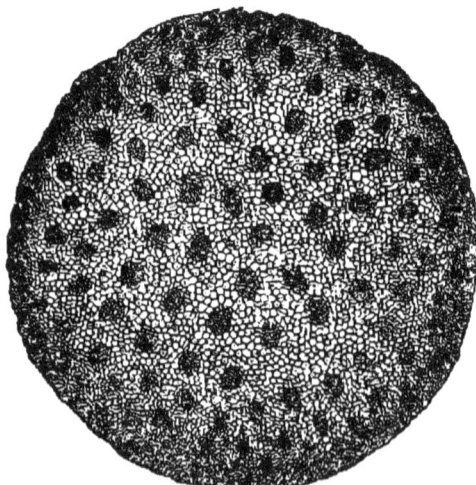

FIG. 14.

In the latter it generally forms a central *pith*, connected with the bark by more or less developed

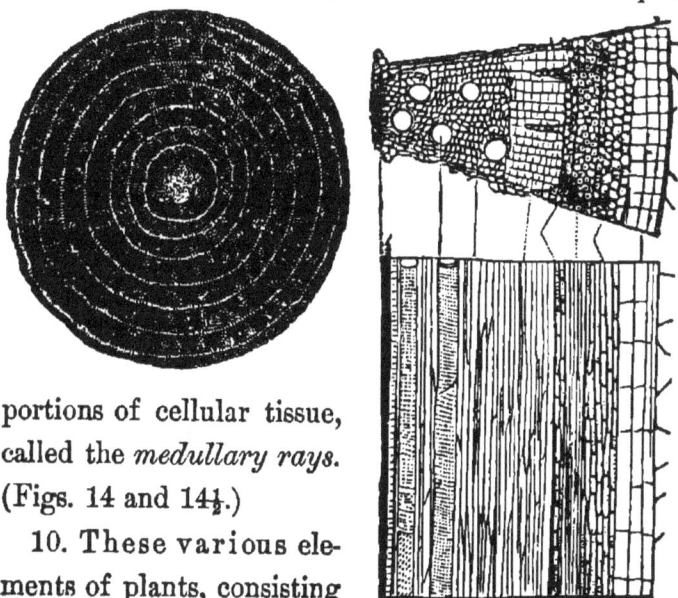

Fig. 14½.

portions of cellular tissue, called the *medullary rays.* (Figs. 14 and 14½.)

10. These various elements of plants, consisting of different forms of cells, tissues, and fibro-vascular bundles, are arranged in each species in a characteristic manner, so that it is often possible to recognize the species from a small fragment of the plant. For this purpose small transparent sections of a stem are prepared, cut in three different ways—transversely, longitudinally through the center, and a section parallel to the last. These are mounted on glass slips, three inches long by one wide, saturated with Canada balsam or other preservative fluid, and covered with very thin glass for microscopic examination.

CHAPTER VII.

PROTOPHYTES, OR THE SIMPLEST FORMS OF PLANTS.

1. THE simplest form of individual plant life is a particle of living matter inclosed in a membrane or cell-wall. Chap. V, Sec. 4. Such plants are called one-celled. Some of these remain entirely distinct from other cells, others form cell-families (Chap. VI, Sec. 1) in a sort of gelatinous investment, while other kinds form a sort of fiber or rod by the adhesion of cells end to end. As each cell in these different kinds is capable of independent life and growth they are all classed as *unicellular*.

2. The green slime which grows on stones or boards in damp places contains many of these one-celled plants. One of the simplest forms is shown in Fig. 15. It is often found in rain-water casks, and is called the *Protococcus*—from

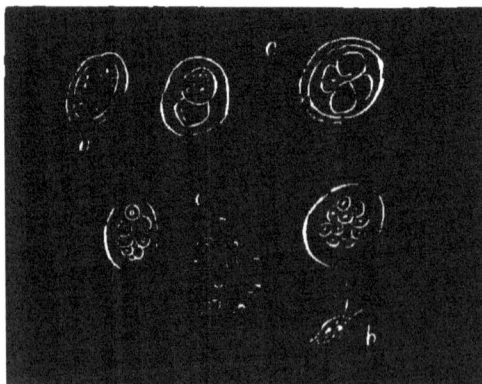

Fig. 15.

two Greek words: *protos*, first, and *coccus*, a berry.
Each cell is round, and varies in color from bright
green to bright red, according to the nature of the
coloring matter diffused in the form of granules
through the bioplasm. It requires a microscope to
see the form and structure of these cells. Each cell
is a perfect plant, and like all vegetables which contain chlorophyll, (Chap. V, Sec. 9,) under the influence
of sunlight, breaks up the carbonic acid gas which it
absorbs from the air, retaining the carbon and giving
off the oxygen. In the dark, however, all plants absorb oxygen and give off carbonic acid, rendering it
unsafe to have many plants in a sleeping-room, since
carbonic acid gas is unfit for respiration by men and
animals.

The cell-wall of the protococcus is quite transparent, and if burst will allow the bioplasm and the
granules of red or green chlorophyll to escape. The
cells multiply by self-division, each one producing
two, four, eight, or sixteen cells. These new cells
differ from the parent cell in not remaining quiet or
still, but having the power of active movement.
They swim about like animals by means of two filaments or *cilia*, (Lat. *cilium*, an eyelash.) These
moving cells may also subdivide into smaller ones
which have been called by botanists *zoospores*, (Gr.
zoos, life, and *sporon*, seed,) or living seeds. Many
of the moving cells, however, lose their cilia and

become stationary. In this state the cell-wall may thicken, and the pond where it dwells dry up, but the mass of bioplasm in the cell may retain dormant life for years, and be ready to resume its work as soon as moisture and warmth shall set it free.

Many of these resting cells, with red chlorophyll, are found occasionally in the snow of northern regions, or near the tops of high mountains, perhaps

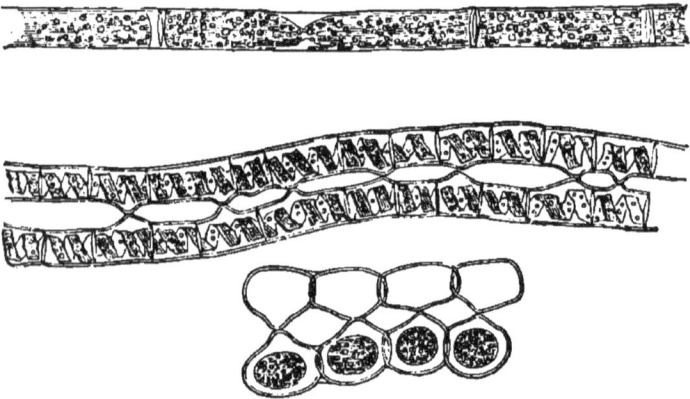

Fig. 16.

carried there by winds, and, finding moisture and sunlight, multiply themselves so rapidly as to color the snow by their multitudes. This forms what is known as "Red Snow." Similar cells may grow rapidly in damp places and form masses which look like coagulated blood. In this way we may account for the so-called showers of blood which have sometimes alarmed the superstitious.

3. Another kind of primitive plants may be found

. as green fibers or threads in almost every running
stream. The microscope will show the cells of these
fibers applied end to end—the cells at each end multiplying by self-division. In another kind, adjacent
cells grow together, and the green chlorophyll and
bioplasm mix in one of the cells, forming a sort of
spore, or seed, which produces a new filament by cell-
division. (Fig. 16.)

4. The unicellular plants most interesting to those
who study with the microscope are called *Diatoms*,
(from two Greek words signifying to cut through,)
because of the ease with which a chain of them may
be broken up into individual cells. These cells contain chlorophyll, generally of a brownish color, and
the external membrane, or cell-wall, is hardened by a
deposit of flinty matter. There are many kinds of
Diatoms, with flinty shells, beautifully marked with
lines and dots, often surpassing the most complicate
patterns of art. Some are globular in shape, some
flat with sides like a pill-box, others square, triangu-
lar, boat-shaped, etc. They move about in the water,
so that some have thought them to be animals. Many
are so minute ·as to require the very finest microscope to make out their details. Each cell consists of
two valves, or plates, applied together like the valves
in a muscle-shell. Fig. 17 shows a valve of one of
the most ornamental diatoms—the *Arachnoidiscus
Ehrenbergii*. The first of these words signifies a disk

built like a spider's web, and is applied to the genus. .
The second word refers to the name of a distinguished

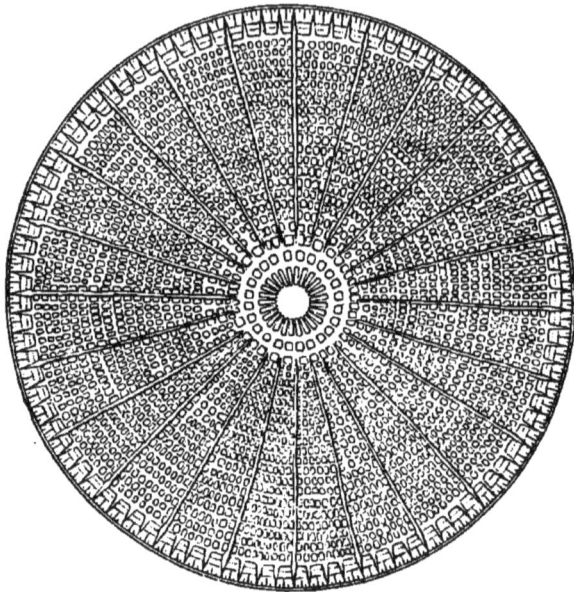

Fig. 17.

naturalist, and names the species—the Arachnoidiscus
of Ehrenberg.   This example illustrates how the various kinds of organisms are named.

In Fig. 18 there are examples of three other kinds
of diatoms.

In the living state diatoms are found abundantly in
every pond, rivulet, ocean, and rock-pool.   They
often make immense deposits by their rapid multiplication, and in a fossil state they form large strata
of rock material.   Under the cities of Richmond and

Petersburg, in Virginia, is such a stratum 20–40 feet thick.

5. Each kind, or species, of unicellular plants, like every other organic form, has endowments or instincts of its own. One sort remains a rounded cell,

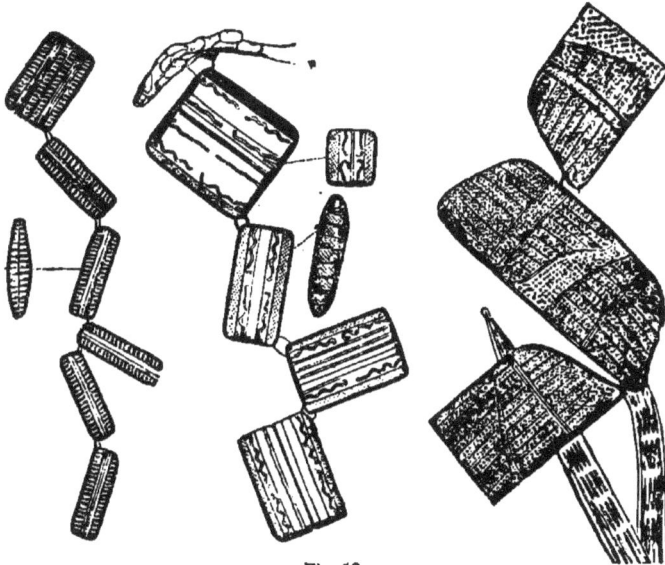

Fig. 18.

or changes to a motile one. Another becomes elongated, and distributes its coloring matter inside in spiral form. Others appropriate glassy flint from their food, and deposit it in beautiful patterns upon their outer cell-wall. Each cell, however, in this type of Protophytes is capable of independent life apart from the rest, and may be considered as a complete individual.

CHAPTER VIII.

THALLOGENS, OR DIVISION OF LABOR IN PLANTS.

1. In the plants which we have been considering each cell is an individual, but in all other kinds the plant, or individual, is made up of many cells, each one of which has a special work to do. Some belong to the root, some to the axis, or stem, and others to the leaves, flowers, or seeds. In the type of Thallogens there is no very accurate division of root, stem, leaves, and flowers, and the whole plant is called a *thallus*, a frond, or green expansion. Under this type we find the classes of *Algæ*, or Sea-weeds; *Lichens*, or the dry, leafy, or mossy patches on trees, stones, etc.; and *Fungi*, or mushrooms, molds, and their allies.

2. The *Algæ*, or Sea-weeds, have been divided into three orders, the Red, the Olive, and the Green sea-weeds. In the more complicate forms we find a sort of distinction of root, stem, and leaf, which reminds us of still higher plants, but the distinction is more apparent than real, since the root and stem serve little other use than the mere mechanical attachment of the plant. The whole plant is made up of cells, and there are no proper vessels. Chap. VI,

Sec. 4. These external resemblances seem uncon-
scious prophecies of forms which prove that all living
things have been formed upon an intelligent plan.

The cells of *algæ* multiply by self-division, and are
of various forms. Some absorb nourishment, or
secrete various materials, or serve merely for growth,
while others are appropriated to the reproduction of
the species, which in some instances has a very com-
plicated method.

3. The class of *Fungi* contains a large number of
different kinds, some quite simple and others complex
in structure. These organisms have so many peculiarities that some scientists regard them as neither
animal nor vegetable, but as forming a sort of third
kingdom. They have a similar cellular form to vegetables, but they have no chlorophyll, as green vegetables have ; light is not necessary to their growth as it is
to vegetables ; and, like animals, they need organic substances for food. It has been found that they are the
agents of fermentation and putrefaction, and their principal business seems to be the removal of the waste material of both animal and vegetable life. They are
universal scavengers. The simplest forms of *Fungi*,
or the *Molds*, resemble Protophytes, except in the ab-
sence of chlorophyll. The yeast-plant is one of these
forms. It is simply a round or oval cell which mul-
tiplies rapidly by putting forth buds and by self-division. It is the cause of fermentation in all sugary

solutions. *Bacteria* of different kinds are minute
fungi, similar to the yeast-plant, but living in solutions of animal matter, in which their growth causes
putrefaction.   Recent studies render it likely that the
more simple forms are but imperfectly-developed
stages in the life-history of other kinds.   Fig. 19
shows the appearance of *bacteria* when greatly mag-
nified.   All *fungi* are made up of elongated cells,
sometimes branching and sometimes membranous or

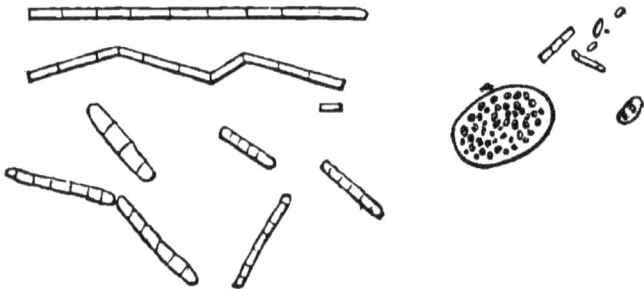

Fig. 19.

pulpy, forming a *mycelium*, or spawn, and rounded
cells forming *spores*, or seeds, which may be supported
on filaments or contained in sacs.   The common
mushroom is one of the larger *fungi*, and the white
or green mold on preserves, cheese, etc., an example
of the minuter kinds.

Many diseases of plants and animals are associated
with the presence of *fungi*.   Mildew and rust in
wheat, the potato-blight, the false membrane in diph-
theria, and many other evidences of diseased action,

seem to depend upon the growth of these parasites. In many cases, however, it is not fully ascertained whether the *bacteria*, or other fungus, is the cause of the diseased condition, or is present, because of the disease, to remove the decaying material.

4

CHAPTER IX.

ACROGENS, OR PLANTS WHICH GROW AT THE SUMMIT.

1. IN fresh-water ponds and rivers, growing in tangled masses of a dull green color, we may often

Fig. 20.

find plants with stems about as thick as a stout needle, but perhaps three or four feet long, with branchlets (improperly called leaves) arranged in whorls at regular intervals upon the axis. These are the stone-worts, consisting of two genera, *Chara* and *Nitella*. In the latter the stem is a simple tube, but in Chara the central or axial cell is surrounded in a spiral manner by others. In Chara, also, the stem is incrusted by carbonate of lime. Fig. 20 will sufficiently illustrate their forms. The points on the axis, or stem, from which the branchlets spring, are called *nodes*, and the intervening parts are the *internodes*.

These stone-worts illustrate the manner of growth in the type of Acrogens. Each internode is formed by the growth and elongation of single cells. The terminal bud is also formed by a single cell, which subdivides into two. One of the latter forms the internode, while the other subdivides into lateral cells, which by continual division produce the branchlets. After a time the terminal cell in the latter is incapable of further division, but in the stem the process continues indefinitely. (Fig. 21.)

These stone-worts are also reproduced by certain organs which grow at certain parts of the axis. These organs are of two kinds, oval *sporangia*, or spore-fruits, and *antheridia*, or organs which contain filaments corresponding to

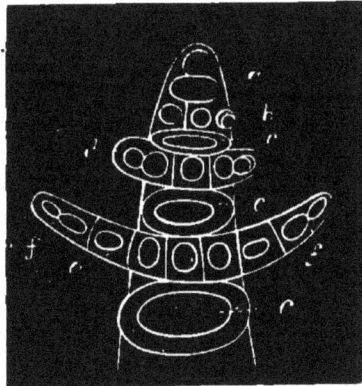

Fig. 21.

the *anthers*, or male organs, of flowering plants. The cells of each antheridium contain little swimming bodies called *antherozoids*, (living anthers.) (Fig. 22.)

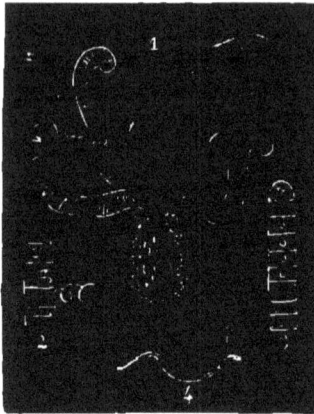

Fig. 22.

These coiled up antherozoids twist and turn about until they escape from the cell and swim in the fluid by means of their two cilia. They find their way to the spore-cases, and by coalescence form the *oospore*, (the egg-spore, or embryo,) from which the future plant is derived. When we consider the power of motion in these organs and others similar to them, we are obliged to admit that sharp lines of distinction between plants and animals are impossible in these apparently simple forms. The growing spore of the stoneworts gives off two filaments, one of which serves as a temporary root, while a cell in the other produces a group of lateral projections from which the young plant springs. This temporary structure is termed the *pro-embryo*, and something similar to this is common to all the Acrogens.

2. *Ferns* form another family of Acrogens, or summit-growers. Their various species are admired for their beautiful *fronds*, often improperly called leaves, and books of collectors often grace the parlor table.

They vary in size, from the Tree-ferns of the tropics, which may be fifty or sixty feet high, to the delicate Maiden-hair fern of the shady dell. In temperate climes ferns have usually a simple or branched under-ground stem, called a *rhizome*, a root-stalk, from which grow root-hairs and fronds. The epidermis of the stem is of brownish hue, and when young and above ground is provided with stomata. Chap. VI, Sec. 6. As in higher plants, the general cellular structure consists of many-sided cells, containing

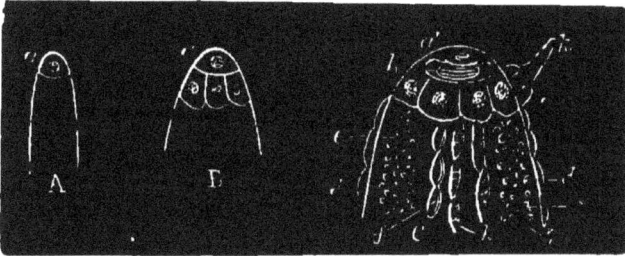

Fig. 23.

chlorophyll and starch granules. There are also vessels, (annular, spiral, and scalariform, or ladder-like,) and fibrous or woody tissue, together forming the harder tissues.

Fig. 23 illustrates the growth of a fern at the summit, together with the metamorphosis of the terminal cell into the various tissues.

In flowering plants the terminal cell of the leaf-bud becomes barren, and the enlargement of the leaf depends on the multiplication and growth of cells

nearer the base, but in the fern the frond grows as
the stem does, so that the peduncle, or stalk, is first
formed, then the embryo frond, then the pinnules, or
wings, etc.

Underneath the frond of a fern we may sometimes
see little brown patches.   Each patch is called a *sorus*,
(plural, *sori*.)   It is sometimes covered by a membrane called an *indusium*, and the little brown bodies
constituting it are spore-cases which have been developed from epidermal cells.   An elastic ring sur-
rounds each spore-case and assists in opening it.   The
growth of the minute spores in the spore-cases may be
watched from time to time under the microscope.
The little spore swells and bursts, and sends out a
rootlet into the soil.   Then a number of delicate cells
are formed from the mother-cell in the spore, making
a little green scale, which throws out rootlets on its
under side.   This *prothallium* (as it is called) produces two kinds of cells, one set which contains spiral
filaments which escape and, by apparently spontaneous movements enter the others, or germ-cells, from
which the future fern is produced.   Fig. 24 gives
a good representation of the various parts in the
structure and life-history of a fern.

3. *Mosses* are also examples of plants which grow
at the tip, or summit.   They are minute and lowly
plants, but are by no means insignificant.   They have
distinct axes of growth, and their delicate leaves are

arranged with great regularity. The stem shows some
indication of the separation of a bark-like portion
from the pith-like, by the intervention of a circle of
bundles of elongated cells, from which branches pass
into the leaves, so as to afford them a sort of midrib.

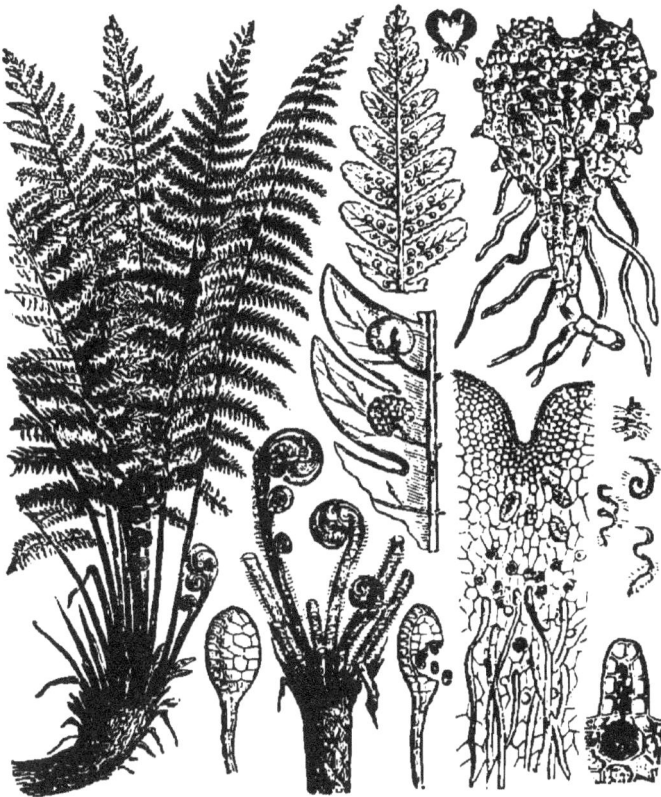

Fig. 24.

The root-fibers are long, tubular cells, quite transparent, within which the circulation of bioplasm may
be seen.

The stems of mosses usually terminate in filaments, each supporting an urn-shaped vessel closed by a lid. The urn is covered by a cap, or hood. Under the lid the edge of the urn has a toothed fringe, and within the urn, or spore-capsule, are double-coated spores. (Fig. 25.)

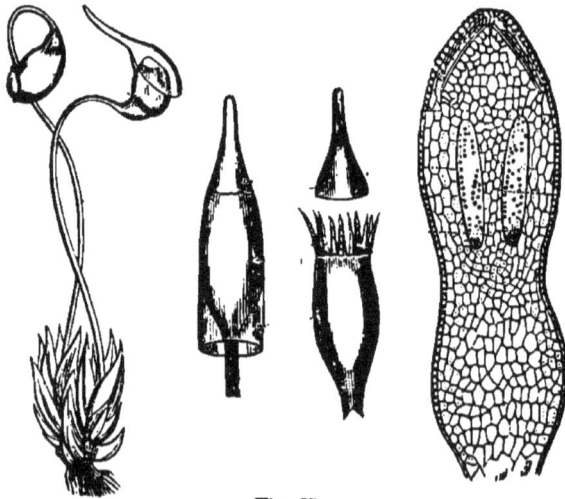

Fig. 25.

In producing new plants, the outer coat of the spore bursts and the inner wall protrudes. New cells grow from the extremity, forming a filament, whose cells at certain points multiply by subdivision, so as to form rounded clusters, from each of which an independent plant may arise.

The minuteness of the spores of mosses and similar plants accounts for their general distribution, even in

the most distant and barren places. Bare rocks raised
from the bottom of the sea, and lava-flows from the
tops of volcanoes, marshes and dry mountain-tops, are
soon covered by mosses and their allies.

CHAPTER X.

ENDOGENS, OR INSIDE-GROWERS.

1. GRASSES, rushes, lilies, and palms, with similar families of plants, are found in the type of Endogens, Chap. IV, Sec. 5. This term was given to them because it was thought that their woody and vascular fibers grew from the inside, and pushed the earlier-formed bundles of fibers toward the circumference of the stem. More exact examinations have shown that the fibro-vascular bundles grow within the cellular or fundamental tissue, turn first inward toward the center or pith, and then bend outward and pass into the leaves. In grasses the cells of the center disappear except at the nodes, (Chap. IX, Sec. 1,) leaving the stem hollow.

2. Endogens are often called *Monocotyledons*, (Greek *monos*, one, and *kotyledon*, a seed-lobe,) because the young plant has but a single seed-lobe. Exogens have two seed-lobes, as we may see in a sprouting bean or pea, and are called *Dicotyledons*, (Gr. *dis*, two.) Acrogens and Thallogens have no seed-lobe at all, but are propagated by cellular spores, and are called *Acotyledons*, (Gr. *a*, without.)

3. In Endogens and Exogens we find a more com-

plete development of the root, stem, and leaves than in the other types, giving a character to the external form of plants which enables us to recognize them and place them in a natural system of classification. It is therefore appropriate here to consider these structures in as brief and comprehensive manner as possible.

4. If a pea or bean be soaked in water, and the leathery skin be stripped off, two large fleshy masses will be seen (the *cotyledons*) inclosing a small cylindrical body, (the *axis*,) which bears two minute leaves at its extremity. The cotyledons and axis together constitute the *embryo*.

In the growing plant the *stem* grows from the axis upward and the *root* downward, and the leaves develop only on the ascending part of the axis and not on the root.

In the growing stem the terminal cells (*a*, Fig. 26, A) multiply and enlarge. They furnish new cells to the cambium layer, or that between the bark and

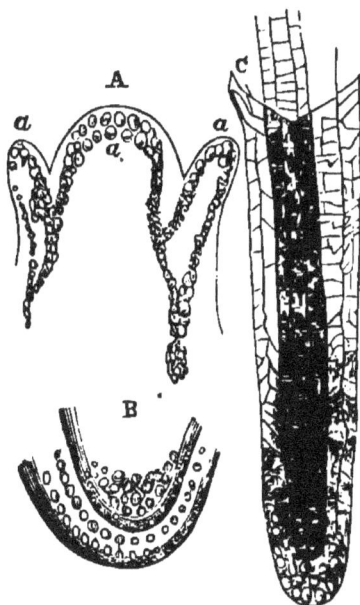

Fig. 26.

wood of Exogens. Chap. VI, Sec. 8. In the root, the multiplying cells are not quite at the extremity. A sort of cap is formed, which receives additions to its interior and pushes out the layers external to them. Thus the new-formed tissue is protected from the rough soil. (Fig. 26, B, C.) Sometimes the absorbing activity of the points of the root-hairs is so great that particles of soil actually unite with them, to be afterward dissolved by the cell-sap.

5. The *primary root* is that which is formed by the downward elongation of the axis. It is in a line with the stem. It is called a *tap-root* when it is thicker than the branches which spring from it, and may be *fusiform* or spindle-shaped like the carrot, *napiform* or turnip-shaped as the radish or turnip, *filiform* or threadlike, or *cylindrical*. *Secondary* or lateral roots are those which spring laterally from the stem or primary root, as the clasping roots of ivy. Sometimes the primary root is undeveloped, or dies, and is replaced by secondary roots. In grasses these are filiform, and are called *fibrous roots*. Sometimes secondary roots become *tuberous* or *fasciculated*, (swollen in the middle, or at intervals,) as in the dahlia, etc. All roots are more or less branched, and are covered with delicate *root-hairs*. If the branches of the root run near the surface of the ground, they are called *creeping roots*.

6. The stem is that part of the plant which bears

the leaves, flowers, and fruit. Some plants are apparently stemless, from this part remaining very short and undeveloped in proportion to the roots and leaves, as in the primrose. The woody stem or *trunk* characterizes trees and shrubs. A simple unbranched trunk, as the palm, is called a *candex*. The *scape* is a leafless stem bearing only flowers, and belonging to a so-called stemless plant. It may bear only a single flower, as the tulip, or several, as the hyacinth and lily of the valley. Sometimes stems send out *runners* or branches which run above the ground, and send out adventitious roots from their nodes or extremities which develop perfect plants, as the strawberry. The *rhizome* is an under-ground stem, sending up branches into the air. The *tuber* is a thickened, fleshy under-ground stem, as the potato, in which we may find buds or *eyes* concealed in depressions. The *bulb* is also fleshy, but has scales surrounding the solid base or *disk* of the stem, or attached to its apex. When the disk is large and surrounded by only a few leaves, as in the crocus, it is called a *corm*. Bulbs may be *squamose* or scaly, *tunicated* as in the onion, *fibrous*, etc.

The length of life of the stem and roots may be only a single year, or *annual;* two years, or *biennial;* or a number of years, or *perennial*.

The trunk or woody stem of Exogens, or ontside-growers, shows on a transverse section a number of circles of fibro-vascular bundles, with cellular rays passing from the pith to the bark. Chap. VI, Secs. 9, 10. These circles are supposed to indicate the layers of annual growth, but this is quite uncertain. Sometimes two or more circles are formed in a year. The diameter and height attained by some Exogens may be very great. The Big-Tree Grove, in Calaveras County, California, contains trees from 350 to 400 feet high, and 33 feet in diameter.

The stems of Endogens, or inside-growers, exhibit in their sections no distinct pith, no concentric circles, no medullary rays, and no separable bark.

7. Stems produce *buds*, which may be regarded as shortened axes, capable of elongation. According to the organs which result from their development, they are termed *stem-buds*, *leaf-buds*, and *flower-buds*. They are *terminal* if produced at the extremity of the primary axis, and *lateral* if at the sides of the axis. In palms and tree-ferns the buds are terminal, and if the top of the stem is cut off the plants perish. Buds are often protected by coarse leaves or *scales*, which may be covered with hairs, or with gummy or resinous matter for additional protection.

Buds often lie dormant, and do not appear as branches unless stimulated by some local injury to the plant; others are altered into thorns. Thorns are undeveloped branches, and many plants which are

thorny when wild are not so under cultivation.
Thorns differ from prickles, which are hardened
hairs.

8. *Leaves* are constituted of cells, with cavities,
fibro-vascular bundles, and epidermis. (Fig. 27.)

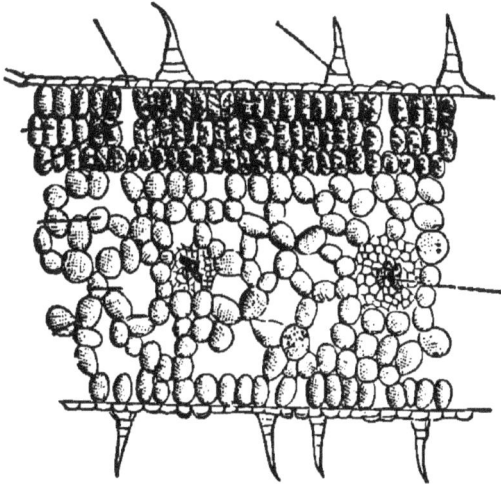

Fig. 27.

The veins in a leaf are the vascular parts, and their
distribution differs in the types of Endogens and
Exogens, so as to afford a ready means of discrimina-
tion. The veins in the leaves of Endogens are generally parallel or straight, and do not form a network
as in Exogens. (Fig. 28.)

When two leaves are at the same level, one on each
side of the stem, they are called *opposite;* when a
circle of leaves is thus produced it is called a *whorl.*
When there is only one leaf on the same level the

leaves are *alternate* or *scattered*. Irregular as the latter mode may appear in different plants, observation shows that it is tolerably uniform in each species. If a spiral line (or thread) is drawn round the stem connecting the points of attachment of the leaves, and these are marked on the spiral, it is found that in any particular species there is a definite number of leaves on any given number of turns made by the spiral round the stem. In the peach and plum the cycle made by the leaves directly above each other embraces five leaves, and the spiral goes twice round the branch. This is expressed by the formula $\frac{2}{5}$. In the alder three leaves form the cycle, and the spiral has but a single turn on the stem. This is represented by the fraction $\frac{1}{3}$.

Covering-leaves are so called because they cover or protect other parts, as the scales of buds, and *bracts*, or leaves in axils of which flowers are placed. The leaf-stalk is called a *petiole*. When it is absent the leaf is said to be *sessile*. At

Fig. 28.

the base of the petiole flat leaf-like appendages are often found, called *stipules*.

Leaves are said to be *simple* when the blade is composed of one piece, however irregular may be its

shape, and *compound* when divided into distinctly
separate parts, or leaflets, connected with the petiole
by secondary petioles.

Leaves may be *lanceolate*, or narrow and tapering;
oblong, or narrow and not tapering; *cordate*, or heart-
shaped; *sagittate*, or arrow-shaped; *ovate*, or egg-
shaped, etc. A compound leaf having leaflets placed
laterally is called *pinnate*. If the leaflets are themselves divided, it is *bipinnate*, and a further division
of the leaflets is *tripinnate*. Sometimes a compound
leaf is triple, or *ternate*, etc. When a ternate leaf
divides twice it is *biternate;* when thrice, *triter-
nate*.

It is not uncommon to find on the same plant leaves
of different forms. The *radical* leaves, or those
which grow from the lower part of the stem, are often
different from the upper ones.

The function, or use, of leaves is to expose the
juices of the plant to light and air, and thus aid in
forming the woody matter of the stem and the various secretions. If the leaves are excluded from air
and light, as is the case in crowded plantations, the
wood is not properly formed. The same may be said
of all the substances formed by the plant. Thus,
potatoes grown in the shade, which impedes the action of the leaves, become watery, and produce little
starch in their tubers.

Leaves also exhale watery fluid, and by decompos-
5

ing carbonic acid gas are able to appropriate the carbon as food, and return the oxygen to the air. Chap. V, Sec. 9.

The length of life of leaves varies greatly. In temperate climates the majority of leaves fall off in the autumn, or are *deciduous*. In the so-called *evergreen* trees and shrubs they persist through the winter, and may even remain several years.

9. The root, stem, and leaves of a plant constitute its organs of nutrition. Fluid matters are taken up from the soil by the cells of the roots, these are conveyed to the leaves, and under the influence of air and light are fitted for the purposes of plant life, and for the production of various materials, as starch, gum, sugar, woody matter, gluten, oils, and resins. Chap. V, Sec. 9.

10. The *flower* is the organ, or assemblage of organs, for the production of the seed. In Endogens and Exogens this structure is conspicuous, and they are hence called flowering plants, to distinguish them from other types.

Fig. 29 illustrates the general structure and arrangement of parts in a flower. The poet Goethe taught that all the various parts of a plant are modifications of leaves. Not that they were originally leaves and were transformed, but that they are formed of the same elements, arranged upon the same plan, and follow the same general laws as leaves. The

parts of a flower are hence called *floral leaves*. These
are usually arranged in four whorls. The outer whorl

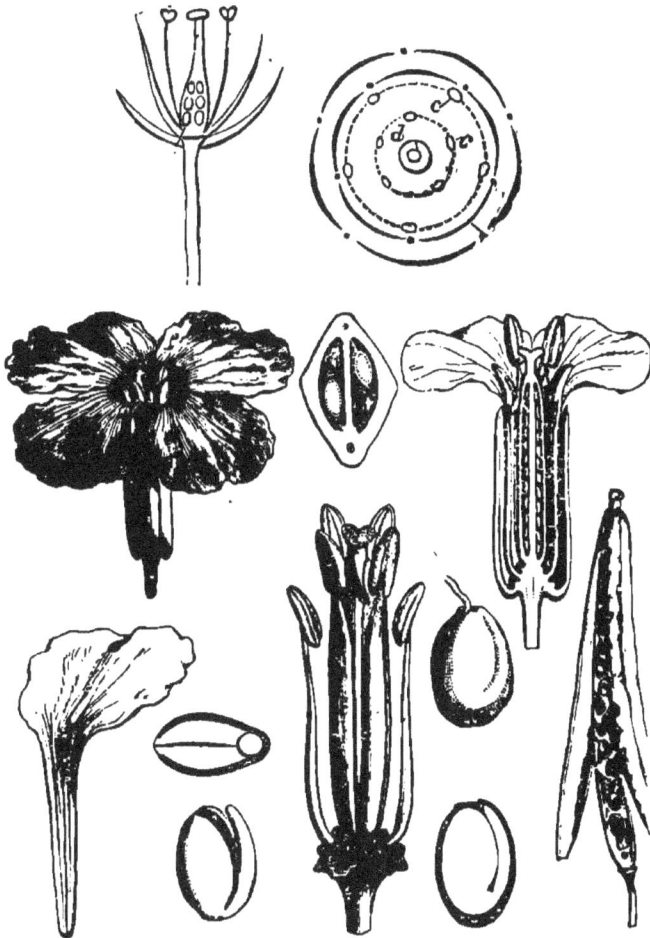

Fig. 29.

is the *calyx*, the next the *corolla*, the third the *stamens*, and the innermost the *pistil*.

The flowers of Exogens exhibit two or five parts, or multiples of these numbers, in their whorls, while Endogens have three, or a multiple of three, in their whorls.   In Exogens, also, the calyx is usually green and the corolla colored, but in Endogens both are often colored.   The term *perianth* is generally applied to the floral envelopes of Endogens.

The parts of the calyx, when separate, are called *sepals*, and the leaves of the corolla *petals*.   Stamens have two parts, the *filament*, or stalk, and the *anther*, or broader portion, corresponding to a folded leaf, and containing fertilizing grains called *pollen*.   The pistil is also made up of two parts, the *ovary*, containing ovules, or young seeds, and the *stigma* for the reception of the pollen-grains.   This latter is sometimes sessile, or resting on the ovary, and sometimes elevated on a stalk, or *style*.

Some flowers have no stamens, and are called female flowers; others have no pistils, and are male flowers.   But these organs are always present, either on the same plant or on different plants.   If the corolla is absent the flower is called *incomplete*, and if corolla and calyx are both absent it is *naked*.   The position of the stamens in relation to the ovary is of botanical importance.   Sometimes they are attached to the *receptacle*, or upper part of the flower-stalk.   They are then below the ovary and free from it, as well as from the calyx, and are said to be *hypogynous*, or under the ovary. Sometimes they are attached to the calyx, but free from the ovary, and are called *perigynous*, or around the ovary. In other cases the stamens appear above the ovary, and are *epigynous*, or upon the ovary. In such instances the calyx is also epigynous. The grains of pollen when discharged from the anther are applied to the stigma, and in a short time send forth tube-like prolongations to the ovule in the ovary, by which means the embryo plant is formed. Many curious and beautiful arrangements are made to ensure the proper application of pollen to the upper part of the pistil. In some flowers the stamens have elastic filaments which are at first bent down and held by the calyx, but when the pollen is ripe the filaments jerk out and scatter the powder on the pistil. The agency of winds and of insects is made use of in some cases. In the hazel, where the pollen is in one set of flowers, the leaves might interfere with the application of pollen, hence they are not produced until it has been scattered.

11. The term *fruit* is applied, in botanical language, to the mature perfect pistil, whether dry or succulent. Fruits are formed in different ways. Some, as the pea and bean, consist solely of the slightly altered pistil; others, as the grape, peach, and plum, have the pistil so changed as to be succulent. The gooseberry, currant, apple, and pear are formed by both pistil and calyx, a portion of the latter

remaining at the top of these fruits in the form of brownish scales. The hazel-fruit consists of the pistil developed into the nut, with a covering of bracts called the husk. The cup of the acorn is also formed by bracts. In the strawberry, the succulent part is the enlarged receptacle, containing numerous small carpels, or fruits, often called seeds. The mulberry, pine-apple, bread-fruit, pine-cone, and fig are made up of numerous pistils formed by separate flowers and combined into a common mass.

12. The *seed* is usually contained in the seed-vessel or fruit. If there is no seed-vessel, as in the fir, the seed is said to be *naked*. In order that the seed may be complete, it must contain the rudiment of the young plant, or *embryo*.

When the seed is placed in favorable circumstances the little plant begins to germinate. Sec. 4. The phenomena of sprouting seed are well seen in the malting of barley. The grain is exposed to moisture, heat, air, and is kept in comparative darkness. These are favorable circumstances analogous to prepared soil. A change takes place in the contents of the grain. The starch, which is insoluble in water, and unfit for the nourishment of the plant, is converted into sugar, which is soluble, and easily absorbed by the bioplasm of the cells as food. The young roots are first protruded, and then the stem, surrounded by a leaf called a cotyledon, or seed-leaf. If the barley were allowed to grow, the whole of the sugar would be used by the plant. But man wishes to get the sugar, and he therefore stops the growth of the plant by drying it, and thus makes malt.

13. *Grasses* and *Sedges* are families of Endogens whose flowers have imbricated bracts, or scales, called *glumes*, instead of a colored perianth. (Fig. 30.)

Among the grasses are classed the nutritious grains, as Wheat, Barley, Oats, Rye, Rice, and Indian Corn.

14. The families of *Palms* and *Bananas* are also noted members of the type of Endogens. (Fig. 31.)

Linnæus, the father of botanical science, called Grasses the plebeians, and Palms the princes,

Fig. 30.

of the vegetable world. The latter are certainly beautiful, and often gigantic, plants. As to utility it would be difficult to make a comparison. Various species of Palms are used for supplying food and for forming habitations. The fruit of some is edible. Many supply oil, wax, starchy matter, and sugar. Their fibers make ropes, and various utensils are formed from their wood or fruit.

15. The *Orchid* family has numerous species, remarkable for the variety of forms and brilliant colors

Fig. 31.

Fig. 32.

in their flowers, which often resemble insects, birds, and lizards. The visits of insects are often needed for their fertilization.

16. The *Lily* family, including many garden flowers, as Tulips, and Lilies, (Fig. 32,) as well as such plants as the Onion, Squill, Aloes, and Asparagus, is a beautiful representative of the type of Endogens. Other families, as that of the *Bulrushes*, have incomplete flowers.

CHAPTER XI.

EXOGENS, OR OUTSIDE GROWERS.

1. PLANTS which produce woody and vascular layers near the circumference of the stem are very numerous, including about 70,000 different species.

Between the woody layers, or rings, and the bark of such plants is a semi-fluid mucilaginous matter containing the new or growing cells. This layer is called the *cambium layer*. At the apex of the stem, and at that of the root, this layer is continuous

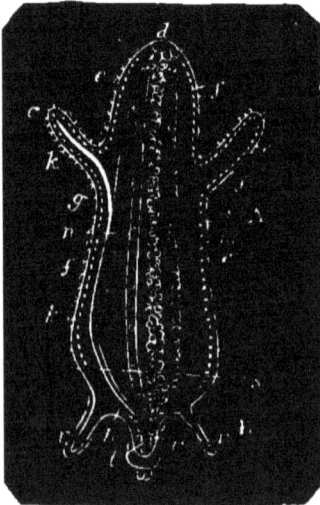

Fig. 33.

with the cells of bioplasm which multiply by self-division in these localities so as to supply the elements of the new tissues. (Fig. 33.)

2. *Incomplete Exogens* are those whose flowers have no corolla. Sometimes, but not always, they have a calyx, or simple perianth. They are of two kinds: 1) Those whose seeds are naked, as in the *Cone-bearing* family, consisting of the Fir and Spruce tribe, the Cypress

tribe, and similar plants. These conifers are generally large trees or evergreen shrubs, and furnish much valuable timber, pitch, turpentine, and resin. (Fig. 34.)

2. Those whose seeds are contained in an ovary, as the Amaranth, Buckwheat, Laurel, Nettle, Fig, and the *Catkin-bearing* family. This latter family is the most important of this order, since it contains the most important timber - trees, as the Alder, Birch, Willow, Poplar, Oak, Chestnut, etc. Their flowers, either male or female, are arranged on a common axis, without separate stalks, and are without either calyx or corolla, but furnished only with scaly bracts. Such clusters, or catkins, at-

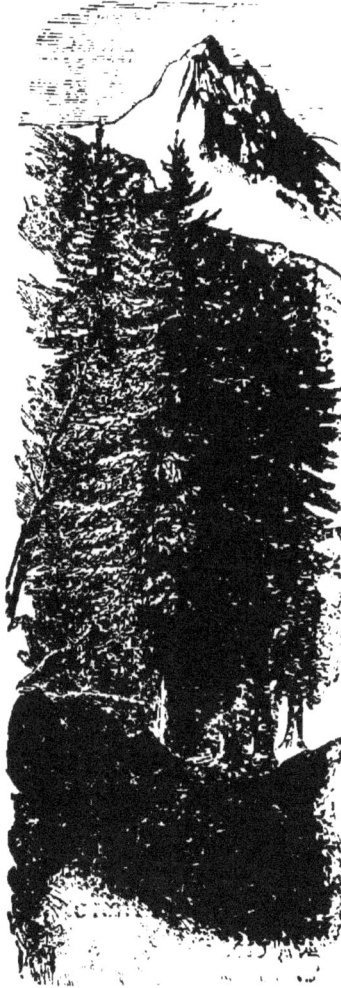

Fig. 34.

tract attention in early spring to the willow, alder, or poplar trees. This family, with the Conifers, gives character to the woodland scenery of temperate climes. (Fig. 35.)

3. In the next subdivision of the type of Exogens we find plants whose flowers have both calyx and corolla. The petals of the corolla are also united, and

Fig. 35.

bear the stamens, as the Honeysuckle, Teazel, Lobelia, Convolvulus, Primrose, Labiate and Composite families, etc.

The *Labiate* family contains many fragrant and aromatic plants, as Mint, Lavender, Sage, Balm, etc. It is characterized by two long and two short stamens,

four little nuts, or naked seeds, and irregular corollas.
The *Composite family* is very extensive. It includes
all such plants as the Thistle, Sunflower, Daisy, Aster,
and Chrysanthemum. There are about twelve thou-
sand species in this family, distributed all over the
globe. They are generally herbaceous plants, and
often contain a milky fluid, or latex. The flowers are
placed on a common expanded receptacle, crowded together into a *capitulum*, or head, and surrounded by
a *general involucre* of densely crowded bracts. The
florets of the central part of the capitulum are often
of a different structure and color from those of the
margin, and the two kinds are distinguished as florets
of the disk and florets of the ray.

4. Another class of Exogens also have calyx and ‐
corolla, but the corolla has distinct petals, and the
stamens are attached to the calyx.

The *Umbelliferous* family is found in this division,
and contains culinary plants, such as Carrot, Celery,
Parsley, and Parsnip ; medicinal herbs, as Caraway,
Fennel, Coriander, and Assafœtida ; and some poison-
ous plants, as Hemlock and Fool's Parsley. This
family is named from the mode of its inflorescence.
An *umbel*, like the capitulum, has the stem terminat-
ing in a number of flowers, but each separate flower
is stalked. The umbel is *simple* when the main stem
or peduncle ends in a number of separate stalked
flowers, as in the Cherry, *compound* when it branches

into a number of secondary umbels, as in the majority of genera in the family of *Umbelliferæ*.

The *Leguminose* family, characterized by the ovary
developing into a pod, (or legume,) is also very extensive. It includes many forms of herbs, shrubs,
and trees. Some have flowers resembling a butterfly,
and hence called papilionaceous, as Clover, Lupins,
Peas, and Beans. Others have irregular flowers which
are not papilionaceous, as the Tamarind-tree and
various species of Senna, or Cassia. In other
cases the flowers are regular, the scales of the calyx
in the bud are *valvate*, or touch only at the edges,
and the stamens are sometimes very numerous, as in
different species of Acacia, and the Mimosa, or Sen-
sitive-plant. The *Rose* family is also a very large
one, and includes not only the roses of our gardens,
but Raspberries, Strawberries, Plums, Apples, Pears,
Cherries, Peaches, Apricots, and Almonds.

The *Cactus* family is also found in this division.
It contains many succulent plants, generally destitute
of leaves whose place is supplied by fleshy stems of
grotesque figures. Some are angular, others are
roundish and covered with stiff spines. They vary
in height from a few inches to twenty or thirty feet.
The flowers are often very showy, varying from pure
white to rich scarlet or purple. In Mexico and
Southern California there are numerous species, some
of gigantic size. (Fig. 36.)

5. In the highest class, or most perfect Exogens, the calyx and corolla are present, the petals are dis-

Fig. 36.

tinct and inserted into the receptacle, and the stamens grow from beneath the ovary.

The *Crowfoot* family, having distinct carpels above numerous stamens, and embracing the *Ranunculus*, or Buttercup, the Larkspur, Aconite, and Peony; the *Poppy* family, having the carpels united into an undivided ovary; the *Cruciferous* family, readily known by their four cruciate petals, and including many flowers and vegetables, as Wallflower, Cabbage, Turnip, Radish, and Mustard; the *Flax* family; the *Tea* family, containing the Camellias and the Tea-plants; the *Orange* family; the *Maple* family; and many others, are found in this group.

CHAPTER XII.

THE VEGETABLE CLOTHING OF THE WORLD.

1. THE beautiful forms of vegetable life, and the peculiarities of different species, give character to the landscape scenery of the world, and the comparison of the different floras (or groups of plants) on the earth's surface will aid us greatly in our biological generalizations.

2. The study of the distribution of plants over the earth is sometimes termed Botanical Geography. It has been greatly promoted by the travels of Humboldt. Standing a few hundred feet below the summit of Chimborazo, he saw an epitome of the vegetation of the globe—a picture of all climates from the tropics to the poles, with their zones or belts of vegetation. Just above him rose the inaccessible summit of snow—a beautiful image of purity set in the cloudless blue of a tropical sky. The only vegetation present was the Red Snow, or a few Thallogens. The volcanic rocks around him were draped with Lichens, a few Mosses, and Alpine flowers. Beneath, the grassgreen slopes, with varied flowers and willowy shrubs, were succeeded by forest belts, and these latter by the tropical luxuriance at the base of the mountain.

3. Each species of plant has its center of distribution at the spot from which it originally sprang. It is not easy, however, to determine these centers, because of plant migration. All plants are not equally capable of migration, or the strongest would have replaced the rest and occupied all the ground. Migration is also hindered by seas, deserts, mountain-chains, and climate, as well as by the existence of other plants and animals.

4. The transitions from one species to another met with in gradually ascending mountain regions are not such as Darwin's theory of natural selection might lead us to expect, nor do they favor any theory of transmutation of one kind into another. Such a mountain-side is the most appropriate place in the world for practically testing such theories. If any transitional forms ever existed between species we may reasonably expect to find them here. But the Alpine species make their appearance, and those of the plains disappear suddenly at particular elevations, and we find no transitional varieties.

5. Climate has much to do with the similarity in the floras of different regions. This holds good in widely separated regions, as seen in the resemblances of the beeches of Japan and the Straits of Magellan, and of the heaths of the Cape of Good Hope and of Western Europe.

6. Griesbach divides the surface of the earth into

6

twenty-four regions of vegetation or natural floras. Each of these is subdivided into zones, and the character of each zone is determined by its elevation above sea-level. A succession of zones is thus obtained until the line of perpetual snow sets a limit to vegetable life. We do not find in nature such definiteness as our classifications and theories imply, yet there is a general similarity between the flora of any part of the earth and that of a mountain-zone of corresponding temperature. Thus, similar and often identical plants occur in the lower zones of the mountain and in the districts (north or south) having an increase of latitude, and this principle continues until the floras of the snow-line and of the arctic regions generally correspond. Yet, notwithstanding similarities, the floras of mountain and of arctic regions show considerable differences.

7. The base of the mountains near the equator, from the sea-level to about 2,000 feet high, forms the zone of palms and bananas. From this to the height of 4,500 feet is the zone of tree-ferns and figs. In India these are covered by many kinds of peppers, and orchids. In the islands of the Southern Ocean the figs are replaced by tree-like *Urticaceæ*, and the valuable cinchona-trees characterize the South American region. The zone of myrtles and laurels comes next, extending to 6,000 feet. The predominant trees are those with thick, shining leaves, as

myrtles, camellias, and magnolias. Acacias and
heaths attain here their highest development, and
evergreen oaks abound. The laurels occur
mainly near the upper limit of this zone, and are
found also in the next, the zone of evergreen trees,
which reaches the height of 9,000 feet. Next is the
zone of trees with deciduous foliage, which extends
to the height of 10,000 feet. In the tropics this is
only seen on elevated plains. From this to 12,500
feet is the zone of conifers, and thence to 15,000 feet
is the zone of rhododendrons. Here lofty trees disappear, and are replaced by luxuriant meadows and
herbs with thick, shining leaves and magnificent flowers, as the rhododendrons and azaleas. The last zone
is that of Alpine herbs, extending to the snow-line.
The plants are chiefly perennial, with woody roots, a
small amount of foliage, and brightly colored flowers.
Nearly all contain resinous and bitter substances.

The Alps and other mountains of temperate climes
have but five or six zones. The zone of fruit trees
rises to about 2,000 feet. The apple and grape ascend
thus high, but the walnut may be found up to 3,000
feet. The woods here consist chiefly of beeches,
alders, pines, and oak. The zone of beeches may
reach to 5,000 feet. The birch, sycamore, hazel, wild
cherry, and many herbs, as the plantain, dandelion,
and chrysanthemum, attain their upper limit here, and
disappear with the beech. At the same time, we

reach the lower limit of the rhododendron, gentian, primrose, etc. The zone of pines comes next, to 6,000 feet. The zone of Alpine herbs extends to the limit of perpetual snow, 9,000 or 10,000 feet. The dwarf willow, a few rhododendrons, the mountain-heath, and a single azalea are all the woody plants of this zone in the Alps. The zone of cryptogams, or the snow region, has only mosses and lichens, and occasionally the "red-snow plant."

GLOSSARY AND INDEX.

A'cid products of plants, 31.

Acotyle'dons, 58: plants without seed-lobes.

Ac'rogens, 26, 50: plants which grow at the summit.

Albu'men, 12: in animals, white of egg and similar material; in vegetables, the nourishing matter in seeds, etc.

Adventi'tious, 61: accidental or additional.

Al'gæ, 46: water-plants belonging to the type of Thallogens.

Al'kaloids, 31: active principles (not acids) of certain plants; as morphia, quinia, etc.

Alter'nate, 64: by turns.

An'ther, 52: the part of the stamen in a flower which contains pollen.

Antherid'ia, 51: the organ in mosses, etc., corresponding to the anther in flowering plants.

Antherozo'id, 52: the moving male element in plants without flowers.

An'nuals, 61: yearly plants.

An'nular, 35: ring-like.

Amœ'boid, 18: resembling the *Amœba*, one of the most primitive animals.

Arachnoidis'cus, 43: a genus of Diatoms, named for its net-like markings on the disk.

Ax'is, 46: the central column of a plant, round which the other parts are arranged.

Bacte'ria, 48: a minute organism, of globular or rod-like form, generally regarded as one of the Fungi.

Bast-fibers, 35: the vessels or fibers of inner bark.

Bi'oplasm, 13, 15: living germinal or formative matter.

Bien'nial, 61, pertaining to two years.

Biol'ogy, 7: the science of living beings.

Bipin'nate, 65: twice pinnate, or having two series of leaflets.

Bot'any, 8: the science of plants.

Bracts, 64: a small leaf, or scale, from the axil of which a flower or its pedicle proceeds.

Bulbs, 61: a fleshy roundish body consisting of scales or imperfectly developed leaves, producing a stem and roots.

Buds, 62: a small protuberance containing the rudiments of future leaves, flowers, or stem.

Bul'rushes, 73: a large kind of rush, growing in water.

Cac'tus, 78: a family of succulent, prickly plants.

Cap'sule, 9: a sort of cup, or seed-pod.

Cam'bium, 38, 59: the glutinous layer between the bark and wood, from which new tissue is made.

Cam'phor-glands, 36: groups of cells secreting camphor.

Carbon'ic acid gas, 31: a gas imbibed by plants for nutrition, the carbon being retained and oxygen given out.

Car'pels, 70: the leaves forming the pistil, sometimes separate and sometimes united into a single ovary.

Cat'kins, 75: a mode of inflorescence so called from its resemblance to a cat's tail.

Capit'ulum, 77: a thick head or cluster of flowers, as a clover-top or dandelion.

Ca'lyx, 68: the outer leaves of a flower, generally green.

Can'dex, 61: the stem of a palm.

Cell, 13: a mass of living matter.

Cell-wall, 27: the outer layer, or membrane, of vegetable cells.

Cel'lulose, 28: the starchy material of which the cell-wall is composed.

Cell-contents, 31: materials within cells.

Cel'lular-plants, 38: a term given to those plants which are destitute of vessels.

Cell-families, 33: clusters proceeding from a single cell.

Centers of plant distribution, 80: the original place of plant growth.

Cha'ra, 51: a genus of Stoneworts.

Chemical elements of bioplasm, 14.

Chemical products of bioplasm, 16.

Chlo'rophyll, 31: the green coloring matter in plants, which sometimes becomes red, brown, etc.

En'dosmose, 15: passage of liquid inward through a membrane or porous partition.

Envi'ronment, 29: surrounding circumstances.

Epider'mis, 36: outer skin.

Epig'ynous, 69: upon the ovary, or pistil.

Ex'ogens, 26: plants whose vessels and fibers grow outside, or between the bark and wood.

Ex'osmose, 15: passage of fluid outward through a membrane.

Evolu'tion, 21: a theory of development of things from simple to complex, by a power within themselves.

Exter'nal forms of plants, 59.

Family, 25: a group within an order in classification.

Fascic'ulated, 60: growing in bunches or bundles.

Fermenta'tion, 47: a chemical change produced by the growth of a fungus, the yeast-plant.

Ferns, 52: a class of Acrogens, generally having spores on the back of the fronds or leaves.

Fi'bro-vas'cular, 38, 58: pertaining to vessels and fibers.

Fil'iform, 60: threadlike.

Fil'ament, 43, 50, 54: the part of a stamen supporting the anther. A thread.

Flo'rets, 77: small flowers.

Flo'ra, 80: a group of flowers belonging to a district or country.

Flow'er-struct'ure, 66, 67.

Forms of cells, 30, 31.

Formed material, 15: material or shapes produced by bioplasm, or living matter.

Form'ative layer, 34: the cambium, or layer where new cells grow.

Frond, 57, 46: the leaves of Ferns.

Fun'gi, 46: a class of Thallogens, embracing the mushrooms and molds.

Fundament'al tissue, 38: the cellular parts of plants.

Func'tions of leaves, 65: the uses served by them.

Fu'siform, 60: spindle-shaped.

Geolog'ic history, 9.

Ge'nus, 25: a group within an order in classification.

Germ-cell, 54: a cell answering to the ovary in flowering plants.

Gen'erating-tissue, 34: the cells which form new tissues.

Germina'tion of seeds, 70.

Glumes, 71: husks, or scales, as in the flowers of grasses and grain.

Grass'es, 58, 71: a large family of plants with simple leaves and jointed stems, as wheat, rye, oats, etc.

Grow'ing-point, 34: the place of growth in Acrogens.

Growth at summit, 52.

Growth of Ferns, 52.

Growth of Ex'ogens, 74.

Groups of cells, 33.

Hairs, 37: epidermic appendages.

Heal'ing-tis'sues, 34.

Heat, effects of, 16.

Hypog'ynous, 68: under the ovary.

Iden'tity of bioplasm, 22.

Iner'tia, 17: the tendency of matter to remain in a state of rest or motion.

Individ'uals, 25: persons or things which cannot be divided without loss of identity.

Incomplete Ex'ogens, 68, 74: those whose flowers have no corolla.

Indu'sium, 54: a membranous covering to the fruit-spot of some Ferns.

Inher'ent mo'tion, 18: movement originating from within.

In'stinct, 45: an inward impulse, or tendency.

In'ternodes, 51: the space between the nodes, or places whence the leaves arise.

Inflores'cence, 77: mode of flowering.

Involu'cre, 77: a whorl or set of bracts around a flower, umbel, or head.

La'biate, 76: plants whose flowers have leaf-like projections

Land'scape scen'ery, 81.

La'tex, 35: fluid, generally milky, contained in special vessels and holding peculiar substances in solution.

Lan'ceolate, 65: oblong and tapering.

Legu'minose, 78 : plants having legumes, or pods.

Leaves, 63.

Lich'ens, 46: a class of Thallogens, often improperly called rock-moss, or tree-moss.

Life, 9 : the power by which organized beings live, or the state of animate existence.

Life his'tories, 8.

Liv'ing matter, 12 : bioplasm.

Lil'ies, 73 : a large family of flowering plants, including lilies, tulips, etc.

Mate'rialism, 10: the dogma that all things consist of matter alone.

Med'ullary rays, 39 : rays of cellular tissue or pith, passing from the center to the bark.

Metamor'phosis, 53 : transformation, or change of form or shape.

Microscop'ic sec'tions, 39.

Mo'tions of bioplasm, 18.

Mo'tile, 45 : having power of self-motion.

Monocotyle'dons, 58 : plants with single seed-lobe.

Moth'er-cells, 32 : the cell from which others originate.

Moss'es, 54 : a class of Acrogens, with plants of small size, and stems with narrow, simple leaves.

Mount'ain-vegetation, 83.

Molds, 47 : minute fungous plants.

Multiplica'tion of cells, 32.

Myco'lium, 48 : the filaments, or spawn, from which a fungus grows.

Names of species, 25.

Na'ked, 68 : without calyx or corolla.

Na'piform, 60 : turnip shaped.

Neb'ulous, 8 : cloudy or misty.

Neb'ular hypoth'esis, 8 : theory of development of worlds from primitive fire-mist.

Nec'taries, 36 : groups of cells which secrete honey.

Nu'cleus, 28 : a concentration of vital power in a cell.

Nucle'olus, 28 : a secondary nucleus.

Nodes, 51 : points on the stem from which the leaves proceed.

Nutri'tion, 20 : the process of nourishment.

O'ospore, 52: the embryo of Acrogens, etc.

Or'chids, 71: a group of plants with very peculiar flowers.

Or'ders, 25: subdivisions of classes.

Organ'ic forms, 45: forms of animals or vegetables.

Organiza'tion, 9, 27: the arrangement of organs or parts.

Or'gans of nutri'tion, 66.

Oil-pas'sages, 36: cavities conveying oil.

O'vary, 68: the lower part of the pistil.

O'vate, 65: egg-shaped.

O'vule, 68: a young seed.

Pab'ulum, 15: food.

Palms, 71: a group of conspicuous tropical plants.

Pedun'cle, 77: the stalk of the flower or fruit.

Per'ianth, 68: the envelope of the flower when caylx and corolla are indistinguishable.

Perig'ynous, 69: around the ovary.

Per'manent tis'sue, 34: bells in a plant incapable of further growth.

Peren'nial, 61: plants which live more than two years.

Pet'iole, 64: the foot-stalk of a leaf.

Pet'al, 68: the leaf of the corolla.

Pin'nate, 65: a compound leaf, with leaflets along a common petiole.

Pis'til, 67: the central organ of a flower.

Papiliona'ceous, 78: butterfly-like.

Pin'nule, 54: a branchlet of a pinnate frond or leaf.

Pol'len, 69: the fecundating dust in the anther of a flower.

Pop'pies, 79: a genus of flowering plants.

Phys'ical forces in bioplasm, 15.

Pith, 39: the central cells in the stem of plants.

Primor'dial u'tricle, 28: the mucilaginous layer within the cell-wall of a cell.

Pro'tococcus, 40: a genus of primitive plants.

Pro'tophytes, 26: the simplest type of plants.

Plant migration, 81.

Pro-em'bryo, 52: a temporary structure in the early life-history of Acrogens. etc.

Pro'toplasm, 13: the physical basis of life, or rather, of living matter.

Prothal'lium, 54: a structure in Ferns similar to the pro-embryo in Stone-worts.

Putrefac'tion, 47: decay of albuminoid matter, produced by the action of Fungi.

Rad'ical, 65 : belonging to the root.

Recep'tacle, 68: the tip of the flower-stalk, sometimes dilated, which supports the organs or florets.

Retic'ulated, 35 : like a net-work.

Red snow, 42 : one of the type of Protophytes.

Re'gions of vegeta'tion, 82.

Reproduc'tion, 20: the multiplication of cells or individuals.

Resem'blances of bioplasm, 14.

Rhizome', 53: an under-ground stem.

Roots, 59.

Ros'es, 78 : a family of flowering plants.

Run'ners, 61: branches of stems which give off roots from their nodes.

Rust in wheat, 48: a species of fungus.

Sea-weeds, 46.

Scales, 62: undeveloped leaves protecting buds, etc.

Scape, 61 : a flower-stem without leaves.

Sag'ittate, 65 : arrow-shaped.

Seeds, 70: the part containing the embryo in flowering plants.

Sedg'es, 71 : a family of plants similar to rushes.

Se'pals, 68 : the leaves of the calyx.

Sieve-tubes, 35: cells forming tubes by sieve-like perforations through adjacent ends.

Ses'sile, 64, 68 : without footstalk.

Scalar'iform, 53 : ladder-like.

Show'ers of blood, 42.

Shapes of leaves, 65.

Soul and life, 10.

Sorus, 54: the fruit-spot on a fern.

Spores, 52: parts analogous to the seeds of flowering plants.

Spore'-cases, 52: the vessels containing the spores in Ferns, etc.

Sporan'gia, 51 : spore-cases.

Spe'cies, 25 : a group of individuals which may have descended from a single pair.

THE END.

www.ingramcontent.com/pod-product-compliance
Lightning Source LLC
Chambersburg PA
CBHW022048210326
41519CB00055B/1195